French Fries and
FIZZY POP!

Rebecca Woodbury, Ph.D., M.Ed.

Gravitas Publications Inc.

French Fries and Fizzy Pop!

Illustrations: Janet Moneymaker

French Fries and Fizzy Pop!
ISBN 978-1-950415-14-4

Published by Gravitas Publications Inc.
www.gravitaspublications.com
www.realscience4kids.com

Photo credits: Cover and Title Page By New Africa, AdobeStock; Above: Fries By ha11ok from Pixabay; Soda By ozmen, AdobeStock; P.3. Vegetables By Anna Pelzer on Unsplash, Pizza By Дарья Яковлева from Pixabay, Meat By Robert Owen-Wahl from Pixabay; P. 5. Fruit By Jonas Kakaroto on Unsplash, Cookies By Denise Johnson on Unsplash, Ice Cream, By Grace Mak on Unsplash; P. 7. Lemon By Varintorn Kantawong from Pixabay, Yogurt By Africa Studio, AdobeStock, Vinegar By eskay lim, Adobe Stock, Pickles By JJAVA, AdobeStock; P. 9. Nuts By Alexa from Pixabay, Chips By PDPics from Pixabay, Bacon By Wright Brand Bacon on Unsplash, Fries By ha11ok from Pixabay; P. 11. Tea By Africa Studio, AdobeStock, Kale, By dasuwan, AdobeStock, Brussels Sprouts By valery121283, AdobeStock, Ginger By Natika, AdobeStock, Rocket By Cosmina, AdobeStock; P. 15 Lemon By Varintorn Kantawong from Pixabay, Ice cream By Grace Mak on Unsplash, Soda By ozmen, AdobeStock, Fries By ha11ok from Pixabay, Tea By Africa Studio, AdobeStock; P. 18 Vinegar By eskay lim, AdobeStock, Fries By ha11ok from Pixabay; P. 19. Pie By Pixel-Shot, AdobeStock, Soda By ozmen, AdobeStock; P. 21. By Voyagerix, AdobeStock

RS4K

Which foods do you like best?

Cheese!

Do you like foods
that are sweet?

Do you like foods
that are sour?

Pickles!

Do you like foods
that are salty?

French
fries!

Do you like foods
that are bitter?

I like
ginger!

Why do foods taste different?

I like cheese.
It tastes salty.

Foods taste different because foods have different types of molecules.

Sour Molecules

Sweet Molecules

Salty Molecules

Bitter Molecules

Both Sweet and Salty Molecules

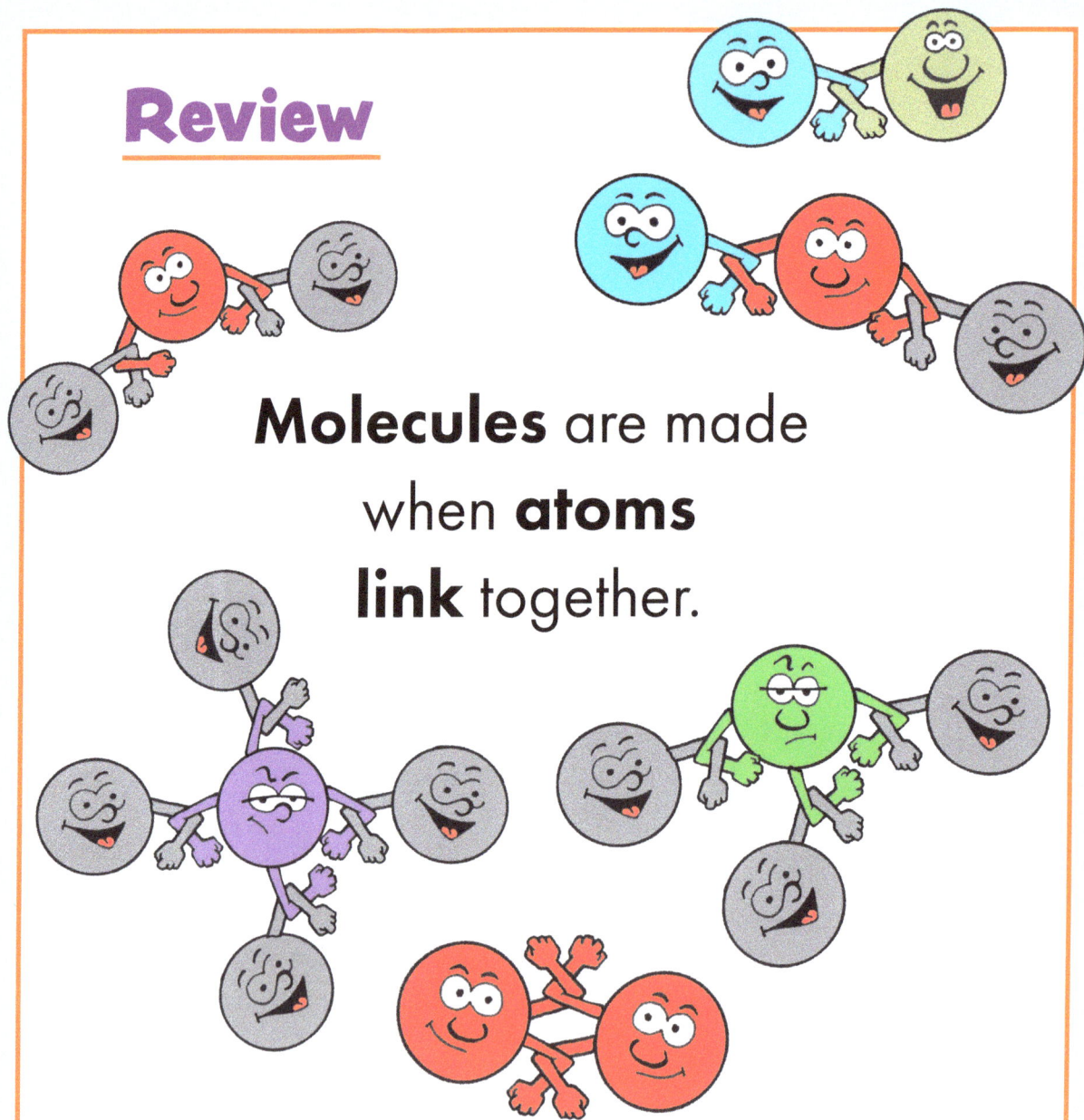

Molecules are made when **atoms** **link** together.

Review

Atoms are tiny building blocks that can link together.

Atoms make everything we see, touch, taste, and smell.

We can draw atoms with arms and hands to help us learn how they work.

Hydrogen   **Oxygen**   **Carbon**

Some tasty molecules.

We atoms make up a sour molecule found in vinegar.

We are a salty molecule found in French fries.

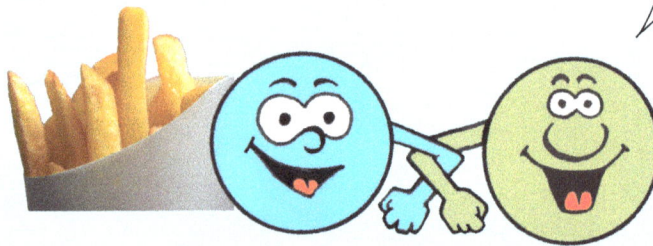

Atoms make up a sweet molecule found in apple pie.

We are a sour molecule found in fizzy soda pop.

You taste all of these molecules with your tongue.

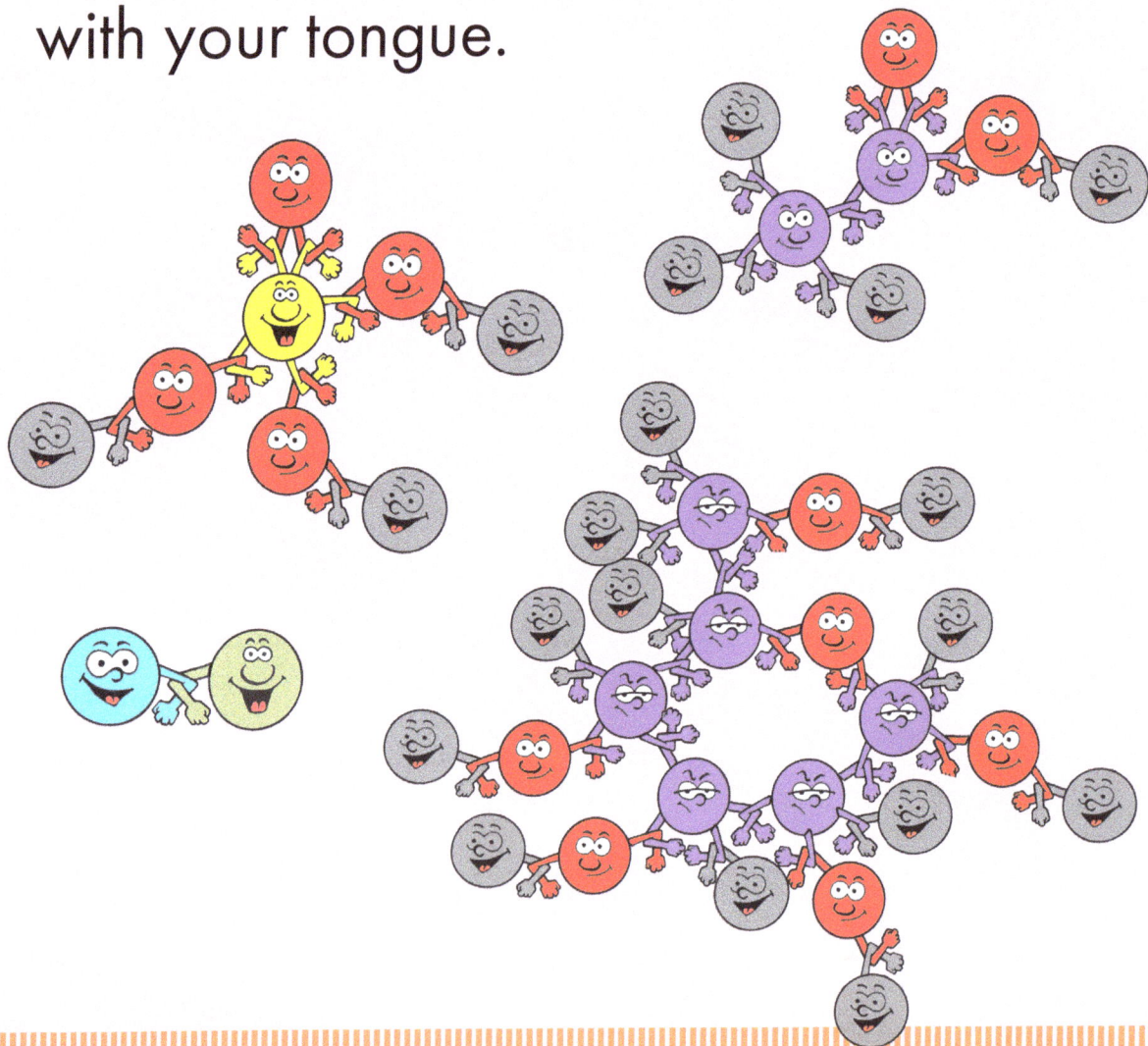

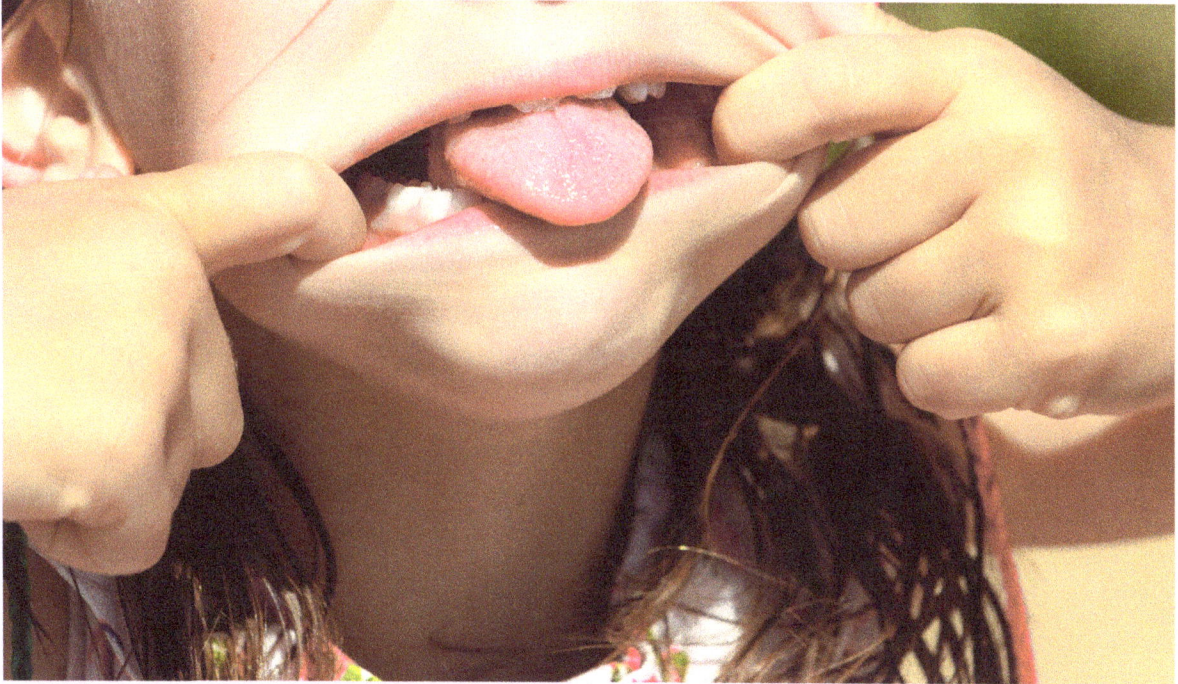

How to say science words

atom   (AA-tum)

bitter   (BIH-tuhr)

molecule   (MAH-lih-kyool)

salty   (SAWL-tee)

science   (SIY-uhns)

sour   (SOW-uhr)

sweet   (SWEET)

tasty   (TAY-stee)

tongue   (TUHNG)

www.ingramcontent.com/pod-product-compliance
Lightning Source LLC
Chambersburg PA
CBHW040150200326
41520CB00028B/7550